EARTH'S CHANGING CLIMATE

UNDERSTANDING GLOBAL ISSUES

Published by Smart Apple Media
1980 Lookout Drive
North Mankato, Minnesota 56003
USA

Library of Congress Cataloging-in-Publication Data
Smith, Trevor.
  Earth's changing climate / Trevor Smith.
    p. cm. -- (Understanding global issues)
Summary: Explores factors in climatic change, such as greenhouse gases
and carbon dioxide emissions, looks at international efforts to remedy
the problem, and discusses whether human activities are contributing to
global warming.
Includes bibliographical references and index.
  ISBN 1-58340-358-2 (lib. bdg. : alk. paper)
 1. Climatic changes--Juvenile literature. 2. Climatic
changes--Environmental aspects--Juvenile literature. [1. Climatic
changes.] I. Title. II. Series.
  QC981.8.C5S65 2004
  363.738'747--dc21

                          2003000022

                     Printed in Malaysia
                     2 4 6 8 9 7 5 3 1

EDITOR Donald Wells **COPY EDITOR** Janice Redlin
TEXT ADAPTATION Trevor Smith **DESIGNER** Terry Paulhus
PHOTO RESEARCHER Wendy Cosh **LAYOUT** Terry Paulhus
SERIES EDITOR Jennifer Nault **CREATIVE COMPANY EDITOR** Jill Weingartz

Contents

Introduction

In 1992, leaders from 108 nations met in Rio de Janeiro, Brazil, to discuss, among other things, **global warming**. They agreed that human activities might be warming Earth more than was normal and that they should take action to reduce the production and **emission** of **greenhouse gases** (GHGs).

The existence of greenhouse gases, such as **carbon dioxide** (CO_2) and **methane** (CH_4), in the **atmosphere** causes Earth to get warmer. It is now known that cars, trucks, buses, factories, production of crops, and other human activities produce a great deal more of these gases than would normally occur in nature.

Since the 1800s, carbon dioxide levels in the atmosphere have risen by 30 percent. The Intergovernmental Panel on Climate Change (IPCC) predicts that carbon dioxide levels will continue to rise and increase between 90 and 250 percent over 1750 levels by the year 2100. It is not easy to reduce GHGs when factories, cars, trucks, and the number of people are increasing. Even a small reduction in human activity would require new ways

Air pollution can cause a variety of serious health problems. Some of these problems include asthma and other respiratory ailments.

to produce power for people, new operating methods for companies that produce GHGs, new ways to travel, and new ways to conserve **fossil fuels**.

Human activities may contribute to global warming.

Since the Rio Earth Summit in 1992, the evidence of global warming has grown. As a result, more people are involved in researching and solving the problem, and private citizens, governments, and corporations have formed groups to improve energy efficiency, save forests and wetlands, and reduce pollution in **developed countries** and **developing countries**. So far, Earth's average temperature has risen between 0.8 and 1.0 °F (0.45–0.6 °C) in the last century. A change of one degree might not seem like much, but it is one of the largest temperature increases in the last 9,000 years.

In spite of this rise in global temperature, people still disagree about whether humans are really capable of changing the temperature of the planet. Hubert Lamb, a respected climatologist, rejects the idea that human "science and modern industry are now

so powerful that any change of climate or the environment must be due to us." Lamb's statement is a reminder that the main factors that influence Earth's climate—the sun, the air, and the oceans—are out of human control.

IPCC, however, has little doubt that human activity can affect the climate. In its *Third Assessment Report*, published in 2001, it states that most of the warming over the last 50 years has been caused by human activities.

These two different opinions on global warming and GHG emissions have created two ways of looking at the world, with large groups of people on either side. One group stresses that the quality of life for all people can be improved only with more development. This group points out that between 1940 and 1970 many climatologists were concerned with another possible ice age as Earth cooled by 0.36 °F (0.2 °C). It has been suggested that this present warming is simply a natural **fluctuation** that is beyond human control.

The other group claims that more development requires the use of more fossil fuels, which would produce more GHGs. They believe that quality of life can be improved without increasing the use of fossil fuels.

The Variability of Climate

Since Earth's formation about 4.6 billion years ago, the atmosphere, sea levels, polar ice caps, global temperatures, and the continents have undergone massive changes. According to paleoclimatologists, there have been periods of intense heat and several ice ages. There were millions of years with no liquid water on the planet's surface, followed by a rainstorm that lasted several million years.

Humans have evolved over the last two million years. During this time, ice ages have occurred regularly, but the ice ages were separated by much warmer periods known as

A glacier's size, history of growth and retreat, and life span all depend on climate conditions.

interglacials. These warmer periods usually lasted 10,000 to 15,000 years, while the ice ages often lasted 10 times as long. All these major climate events occurred before human activity added large amounts of GHGs, such as carbon dioxide, to the atmosphere.

Since emerging from the last Ice Age about 10,000 years ago, Earth has been in a warm interglacial period (called the Holocene). The last 9,000 years have been a period of relative

---

The climate on Earth has been stable for 9,000 years.

---

stability. During this time, Earth's average temperature has remained close to 60 °F (15 °C), rising or falling by no more than one degree. Compare this stable temperature with the "Younger Dryas Event" about 12,000 years ago. During the Younger Dryas Event, temperatures dropped by about 12.5 °F (7 °C) within a few decades.

Even though Earth currently has a relatively stable climate, there are still many major **weather** events that continue to make human history interesting.

GLOBAL WARMING

Since 1880, Earth's temperature has risen about 1 °F (0.6 °C). There have been small swings in global temperature (about 0.2 to 0.4 °F, or around 0.2 °C) in every decade. In some decades, Earth's temperature has risen. In other decades, Earth's temperature has dropped. Over the last century, though, the temperature has moved gradually upward.

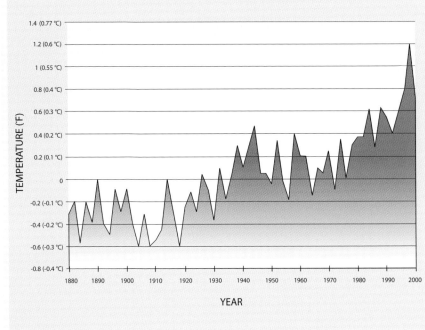

Global Temperature Changes (1880–2000)

These weather events include heavy rainfalls, floods, droughts, tornadoes, and hurricanes. In 1998, 10 to 20 inches (25–50 cm) of rain fell on Texas in the United States. There have been snowfalls in Mexico and unusually high temperatures in Alaska. Although certain regions have experienced unusually high temperatures or large amounts of rainfall in recent years, events such as these will always occur locally within a stable global climate.

While the coming of a new ice age was the main concern in the 1970s, when the temperature of Earth was dropping, global warming has been the major topic of discussion since the 1990s. Already, sea levels have risen six to eight inches (10–20 cm) as the polar ice cap melts in the Arctic. This has caused **oceanographers** and people who live in coastal areas to worry. A significant rise in ocean levels would submerge many coastal areas, and coastal areas tend to be heavily populated. For example, Los Angeles, Miami, and New York City are some of the most heavily populated areas in the United States. Heavy flooding in these areas would be disastrous to society and industry.

Scientists believe that changes in climate cause an increase in thunderstorm and lightning activity.

In light of Earth's climate history, the last 9,000 years of stability do not ensure another 9,000 years of the same climate. Even if it was proven that humans are causing increased global warming, the next wave of ice and snow could come at any time. The recent increase in the global temperature, however, suggests that humans can have an effect on the planet's climate and weather systems.

KEY CONCEPTS

Climate Climate is a description of the weather conditions that typically occur in a large region of the planet. This description can include average temperature, winds, average rain and snowfall (precipitation), and other weather trends. For instance, one can say that the climate in Alaska is cold. Climate does not describe conditions such as temperature or wind speed on a particular day. Weather is a description of the exact temperature, amount of precipitation (if any), and wind speed for a certain day. Climate can describe very long-term trends (for instance, millions of years), but weather changes from day to day.

Glacier A glacier is a buildup of ice, air, water, and rock debris. Glaciers flow very slowly. A glacier can be as large as a continent, such as the ice sheet covering Antarctica, or it can fill a small valley between two mountains, called a valley glacier. The term "glacier" is used to describe ice sheets, ice caps, ice streams, and ice shelves.

Ice age The term "ice age" is used to describe a long period of cold marked by the worldwide buildup of ice and snow. An ice age can be a cool period tens to hundreds of millions of years long, or it can be a period of tens of thousands of years when glaciers are at their peak. When the term is capitalized as "Ice Age," it refers to the last time glaciers were at their peak on Earth. More than 20 glacial advances and retreats have occurred during the last two million years. Ice ages are caused by Earth's changing orbit in relation to the sun, as well as the movement of continents and Earth's tilt.

Paleoclimatologists Paleoclimatologists study past climates and use this information as a guide to what might happen to Earth's climate in the future. Paleoclimatologists look at tree rings, ice cores, pollen residue, and rock sediment. From chemicals found in fossil samples, they can determine the air temperature an organism experienced when alive. An ice core, often more than one mile (1.6 km) long, has bubbles of prehistoric air that can reveal what the climate was like for the past 150,000 years or more.

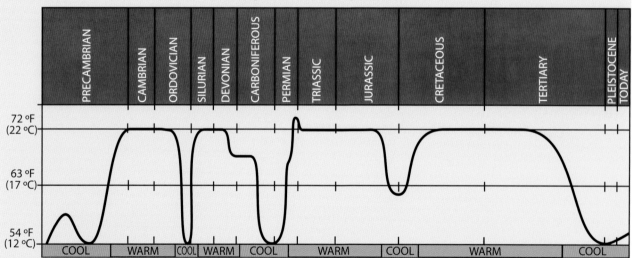

Flooding causes millions of dollars in damage to homes and businesses around the world.

LONG-TERM CLIMATE CHANGE

Earth has experienced major climate changes during its history. It has been cool during the last few periods on Earth, but for the last 4.6 billion years, there have been great fluctuations in temperature.

Average Global Temperature

PRECAMBRIAN
CAMBRIAN
ORDOVICIAN
SILURIAN
DEVONIAN
CARBONIFEROUS
PERMIAN
TRIASSIC
JURASSIC
CRETACEOUS
TERTIARY
PLEISTOCENE
TODAY

72 °F (22 °C)

63 °F (17 °C)

54 °F (12 °C)

COOL   WARM   COOL WARM   COOL   WARM   COOL   WARM   COOL

Climatologist

Duties: Studies long-term trends in climate
Education: Bachelor's degree in climatology, although studies in all natural sciences can be an asset
Interests: Investigation and problem solving, adventure

Navigate to the Web site **www.greenbiz.com** for further information about a career in climatology. Also click on **http://stommel.tamu.edu/~baum/climatology.html** to learn more about climatology.

Careers in Focus

Climatologists are people who like to solve problems. They study how climate is affecting and affected by land use, plant and animal species, society, and other factors. Climatology is one of the more adventurous environmental sciences. A climatologist might drill holes in Antarctic ice, travel to the bottom of the ocean, or journey to the tops of mountains to obtain data. Climatologists might work with marine animals, fitting them with sensors, or take samples of **plankton**, fish, and insects. Some might maintain sensor buoys out in the middle of the oceans. Others will work with complex computer programs and some of the largest, most powerful and expensive computers. No matter how they go about it, the climatologist's goal is to try to predict how Earth's climate will change in the future or determine how it changed in the past.

A climatologist is a scientist who often has a background in any number of the sciences, including physics, meteorology, biology, zoology, botany, paleontology, geology, entomology, microbiology, oceanography, astronomy, math, and computer science. There are also paleoclimatologists who study ice cores, rock samples, and fossils to determine what climate conditions existed in the past.

The Climate System

Earth's climate is a complex group of systems that all affect each other. The interaction of these systems makes it almost impossible to predict climate and weather. Computer models have helped the process of prediction in recent decades. Meteorologists use huge amounts of data, which include temperature, wind, precipitation, and humidity, to generate realistic computer simulations of climates. After a few days, however, the weather system created by a computer begins to look quite different from the real weather. This is why the long-range weather forecast is often imprecise. The real weather situations on the planet are too complex to make an accurate model that holds true into the future.

Still, the major factors that influence climate are known—solar radiation, oceans, atmosphere, and land. The way they interact, however, is complicated by

Tropical rainforests help control Earth's environment by regulating rainfall, reducing soil erosion, and controlling temperatures.

many variables, such as the direction, angle, and speed of Earth's orbit around the sun.

Plants and animals can also affect climate. For example, an area that is covered in trees is dark and absorbs more heat, making that area warmer through a process called the albedo effect. Also, plants and some plankton take carbon dioxide out of the air and convert it to oxygen. Animals remove oxygen from the air and convert it to carbon dioxide.

Plants and animals can influence the greenhouse effect of carbon dioxide. For example, if all of Earth's plants die, carbon dioxide levels will increase. If the plants increase in number, the carbon dioxide levels will fall, and oxygen will increase.

EARTH'S ENERGY BUDGET

Earth intercepts less than one-billionth of the sun's radiation. Radiation comes to Earth as visible light. Of all the sunlight that comes to Earth, only half is absorbed by the land. The rest is reflected back into space or absorbed by the atmosphere and the clouds. The light that is absorbed by the land is converted into infrared radiation or heat. The heat can then return to space or be absorbed by the atmosphere and bounced back to the land. Greenhouse gases trap this heat and bounce it back to Earth.

Solar Energy and Reflection

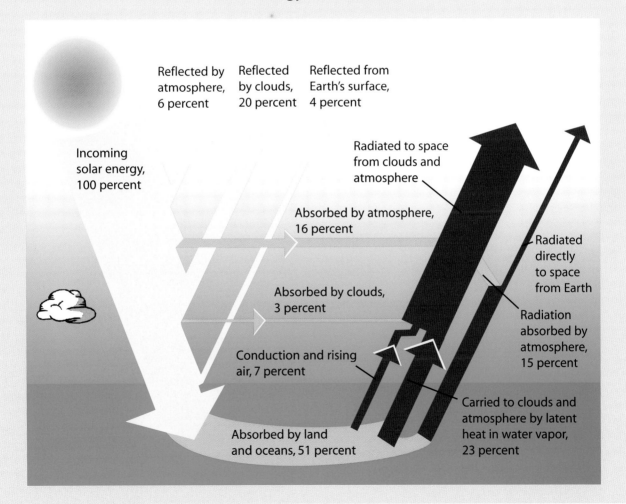

Reflected by atmosphere, 6 percent

Reflected by clouds, 20 percent

Reflected from Earth's surface, 4 percent

Incoming solar energy, 100 percent

Radiated to space from clouds and atmosphere

Absorbed by atmosphere, 16 percent

Radiated directly to space from Earth

Absorbed by clouds, 3 percent

Radiation absorbed by atmosphere, 15 percent

Conduction and rising air, 7 percent

Carried to clouds and atmosphere by latent heat in water vapor, 23 percent

Absorbed by land and oceans, 51 percent

Recently, some research has shown a link between activity on the sun and Earth's climate. For example, high sunspot activity seems to cause a higher average temperature on Earth. A period of low sunspot activity in the 17th century coincided with the coldest period of what became known as the Little Ice Age in Europe (between the 14th and 19th centuries). However, it is still not known for certain if activity on the sun affects the climate of Earth.

Sunspots are dark areas on the surface of the sun. They appear dark because they are cooler than the rest of the sun's visible surface.

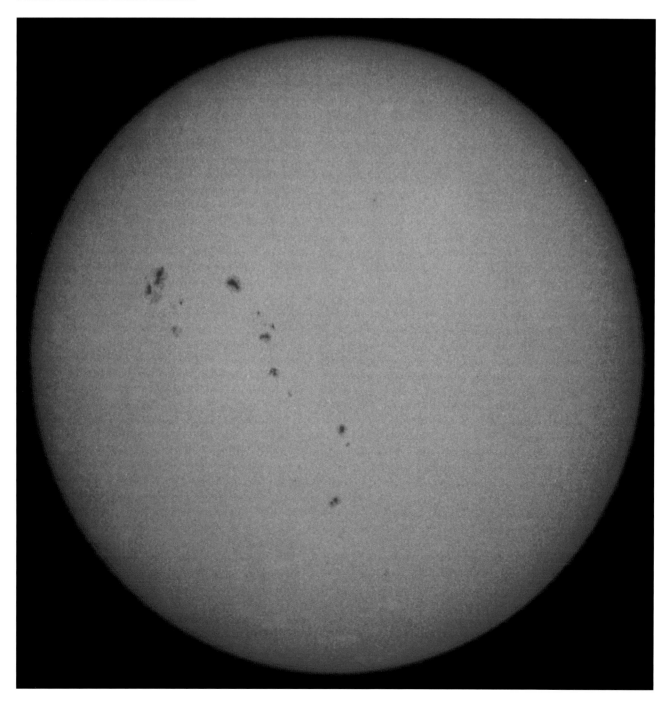

Wind is one system that is affected by the sun. Most of the winds that affect climate are in the weather layer, which is six to nine miles (10–15 km) above Earth's surface. Air circulates as it heats and cools, forming rotating cells of wind. The spinning of planet Earth causes these wind cells to spiral around it. This is called the Coriolis effect. As winds pass over land and water, they cause evaporation, gathering moisture and forming clouds. The effect of clouds on climate is one of the least understood branches of climatology. Some clouds reflect light and cool the planet. Other clouds trap heat and warm Earth.

Climate is complicated by **feedback** systems. There are two types of feedback: positive and negative. For example, when the snow cover begins to melt in the spring, the ground is able to absorb more heat, which increases the rate of melting. This is an example of a negative feedback system in action. A positive feedback system occurs when snow and ice form in winter and reflect heat back into the atmosphere, which allows more ice and snow to form. Feedback systems occur globally and within **biomes**, ecosystems, communities, populations, and organisms.

The time lag involved with some changes can also complicate climate. Sometimes change happens quickly. The atmosphere reacts to the heat of the sun in a very short time,

CLIMATE AND THE SEA

Oceans cover more than 70 percent of Earth, and they play a large role in climate. Oceans store massive amounts of carbon dioxide, heat, and moisture. Most of Earth's clouds, rain, and snow come from evaporation from the oceans. Oceans also absorb heat from hot areas, thus cooling them (as well as delivering warmth to colder areas). Without the warming **Gulf Stream**, western Europe would be sub-arctic. As climate changes, areas closer to water experience fewer changes because the oceans moderate their climate. However, ocean patterns and currents may change as well, which can lead to unpredictable results, such as typhoons.

The Isles of Scilly are located off the southwest coast of Great Britain. These islands experience extremely mild temperatures.

with air temperature and air pressure changing within a matter of hours. At the other extreme, it can take thousands of years to change the deep

---

Oceans stabilize the climate on Earth.

---

ocean's temperature. As a result, this huge amount of water stabilizes Earth's climate, absorbing excess heat or adding heat to a cooling atmosphere. In addition, the oceans absorb and release large amounts of carbon dioxide, much like forests.

Oceans have regular circulations and currents that respond to and affect climate. Their movement depends on wind, the salt and heat content of the water, the surface of the seabed, and Earth's rotation. The most important ocean current is the aptly named conveyor belt, which exchanges seawater between the Pacific and Atlantic Oceans. This current is set in motion when water in the North Atlantic sinks, moves south, circulates around Antarctica, and then moves northward to the Indian, Pacific, and Atlantic basins. It takes 800 years for water from the North Atlantic to reach the North Pacific.

Although the oceans do affect climate, the atmosphere has the greatest effect on global warming. The atmosphere naturally contains 750 billion tons (680 billion t) of carbon, mostly in the form of carbon dioxide. Deforestation and burning fossil fuels add seven billion tons (635 billion t) of carbon dioxide each year. Some of this carbon dioxide is absorbed by the remaining forests and the oceans. Some of it seems to disappear, which is one of climatology's greatest mysteries. The rest of the gas remains in the atmosphere. Carbon dioxide, along with the other human-produced GHGs, such as methane, nitrous oxide, and ozone, represents only about 0.1 percent of the volume of the atmosphere, but it has a big influence on the amount of heat that is radiated back to the planet. Water vapor, which naturally comprises about one percent of the atmosphere, is the most potent greenhouse gas, but the Intergovernmental Panel on Climate Change (IPCC) feels that the gases caused by human activity are the ones that have the most influence on global warming.

KEY CONCEPTS

Albedo effect The albedo effect has to do with the percentage of radiation reflected from a surface compared to the radiation striking it. Main factors affecting a rate of albedo are texture and color. Darker colors absorb more of the sun's rays, while lighter colors reflect the rays. A perfectly reflective surface would have an albedo of 100. Earth has an albedo of about 30. Snow has a higher albedo than grasslands or forests.

Coriolis effect Earth's rotation causes air and ocean currents to be pushed in different directions. This is called the Coriolis effect. In the Northern Hemisphere, air and ocean currents are pushed eastward. In the Southern Hemisphere, they are pushed westward. The Coriolis effect is strongest at the North and South Poles. It is also partially responsible for the movement of some wind cells.

Feedback systems Feedback systems occur on every scale: globally and within biomes, ecosystems, communities, populations, and organisms. Negative feedback is a way for a system to maintain stability. For example, Earth's climate is said to be in balance, but if radiation entering Earth's atmosphere increases, the global temperature rises, and the balance is upset. A negative feedback due to increased radiation would be cloud formation. As Earth warms, the rate of evaporation from the oceans increases and enhances cloud formation. With more cloud cover, more radiation is reflected into space, reducing the amount of radiation entering the atmosphere and lowering the global temperature.

Radiation Radiation is energy that spreads out as it travels. Light from a lamp, microwaves, and radio waves are types of radiation. Radio waves are very low-energy forms of radiation. They have long wavelengths. Microwaves have shorter wavelengths and are more intense. There is also infrared radiation, which is called heat. Earth gives off infrared radiation when struck by visible light. The visible light spectrum is intense, but ultraviolet and gamma radiation are the most intense forms of radiation.

Sunspots and solar wind The sun has regular cycles and a climate. Sunspots occur in magnetically turbulent areas of the sun. They last one or two weeks and reach peak activity about every 11 years. Solar wind is the rippling movement of charged particles that flow out of the sun as it burns. Earth is protected from these winds by its magnetosphere (a magnetic blanket that surrounds Earth).

Duties: Studies factors that affect weather and makes short-term predictions about weather
Education: Bachelor's degree in meteorology or closely related field with much meteorological coursework
Interests: Data collection, research techniques, problem solving

Navigate to the Web site **http://virtualskies.arc.nasa.gov** for more information about a career as a meteorologist. Also click on **www.dss.ucar.edu/other_resources** to learn more about meteorology.

Careers in Focus

Weather people are correctly called meteorologists, but not all meteorologists are weather people on newscasts. A meteorologist can be involved in many tasks and processes. A career in meteorology can lead to research and work in air pollution control, agriculture, air and sea transportation, defense, and the study of trends in Earth's climate, such as global warming or ozone depletion.

Meteorologists who forecast weather are the largest group. They study information about air pressure, temperature, humidity, and wind velocity. Then, using math and computer models, they make short- and long-range weather forecasts.

These forecasts inform not only the general public, but also those who need accurate weather information for both economic and safety reasons: for example, those working in the shipping, aviation, agriculture, fishing, and utilities industries.

Meteorologists advise air traffic control and other agencies about weather hazards such as thunderstorms, developing storm cells and fronts, turbulence, tornadoes, icing, flooding, flash flooding, and other types of weather-related events. They issue weather advisories for vehicles, aircraft, and watercraft to various governmental agencies and the public. They use sophisticated computer software programs that assist them in modeling the potential flow and intensity of storm cells and fronts. They are also available to participate in weather-related research projects that seek to provide more accurate forecasting methods over a longer time period.

Evidence of Global Warming

Although the weather on Earth can change quickly, certain conditions are more or less constant and taken for granted. Plants and animals have evolved to suit local weather conditions that have remained, broadly speaking, the same for centuries. Agriculture and housing have been adapted to climates that have remained the same for long periods of time. However, a study of past climates shows that the relative stability of the last 9,000 years is unusual.

Average global temperature has changed only slightly in the last 9,000 years. Yet, a cooling of 7 °F (4 °C) would take the planet back to the conditions of the last Ice Age. A warming of 7 °F (4 °C) would recreate the warm and wet climate of the Cretaceous Period, when dinosaurs ruled the world.

Although there is concern about global warming, the greenhouse effect keeps Earth's climate stable. Earth receives heat from the sun in the form of sunlight. This is absorbed by the surface of the planet and radiated back into space as heat. It is the blanket of gases, which includes water vapor, carbon dioxide, and other GHGs, that radiates the heat back to Earth, further warming the planet's surface. Without these gases, life on Earth would not exist. This natural process keeps Earth 60 °F (33 °C) warmer than it would be without these gases. The moon, at the same distance from the sun but without this atmospheric blanket, has a temperature range of –279 °F on the dark side to 261 °F on the sun side (–173–127 °C).

Polar bears are drastically affected by climate change. With less food available due to shorter winters and less ice cover, polar bears fail to reproduce strong, healthy offspring. The decrease in this animal's population is just one of many signs of climate change.

LEVELS OF ATMOSPHERIC CARBON DIOXIDE

This chart, which comes from the Climate Modeling and Diagnostics Laboratory of the U.S. Department of Commerce, shows the levels of atmospheric carbon dioxide over the last 30 years. The information comes from four stations (Alaska, Hawai'i, American Samoa, and the South Pole). Compare this chart to the temperature graph shown on page seven. There are similarities, but it does not necessarily mean the two variables are related or that one is affecting the other. However, many people believe the two graphs are related and that the rise in carbon dioxide levels (as well as the other GHGs) is causing global warming.

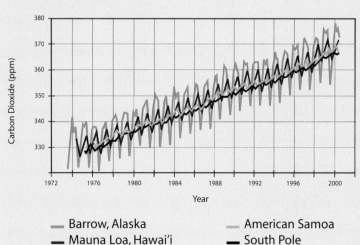

— Barrow, Alaska
— Mauna Loa, Hawai'i
— American Samoa
— South Pole
(ppm = parts per million)

The Intergovernmental Panel on Climate Change (IPCC) declared in 1990 that humans were causing increased greenhouse warming that could raise global temperatures by 2.7 to 8.1 °F (1.5–4.5 °C) by 2050. This news alarmed the global community. What would be the human cost of a global temperature change that would approach the climate of the dinosaurs? What could be done to prevent it, and what could be done to prepare for it? Since that report, IPCC's numbers have changed twice. In 1995, the projection was a 1.8 to 6.3 °F (1.0–3.5 °C) rise in temperatures by 2100. By 2001, the projection was a 2.5 to 10.4 °F (1.4–5.8 °C) rise in temperatures by 2100. These reports are alarming, but how seriously should the world treat such changing predictions?

Meteorologists use observed data and computer models to forecast weather. IPCC uses general circulation models to make its predictions. These are three-dimensional models that divide the globe into thousands of stacked cubes 186 miles across and 0.6 miles high (300 km by 1 km). For each cube, temperature, humidity, wind speed, and pressure are recorded and fed into a powerful computer. The computer then uses this information to predict climate in the future. The problem is that many factors that affect climate are not included in these models. For example, **aerosols**

GLOBAL WARMING AND THE ARCTIC

Research shows that polar bears have been losing weight over the last 30 years. Many are 175 to 200 pounds (80–90 kg) lighter than they would have been 15 years ago. Bears hunt seals from ice that extends over water in the Arctic. The females spend the winter putting on as much fat as possible so they can bear cubs. As temperatures have risen, the ice has started to recede earlier every year. This means the polar bear's hunting grounds have become inaccessible. The bears have to fast through summers that are becoming longer, and they are losing fat stored during the winter. Less fat on their bodies means fewer cubs are born, and those that are born are thin and weak. The polar bear is an example of a species that may not be able to cope with a rapid climate change.

The Inuit who live in the Arctic claim that winters are less fierce than they used to be, and the summers are warmer. In fact, the freeze that occurs in autumn now comes one month late, and spring thaws are coming earlier. Already, the foundations of houses are cracking in some areas. Many fear that one day houses on frozen mud will slide into adjacent lakes.

and other chemicals in the atmosphere that reflect sunlight are not taken into account. It is also difficult to combine

---

Meteorologists use observed data and computer models to forecast weather.

---

atmospheric models with ocean models, and many argue that the cubes IPCC uses are too big to generate accurate predictions.

Some people argue that there is no point in disrupting economic growth based on a prediction that has not been supported by all the experts. Indeed, many scientists believe that another 20 years of study is

needed before a solid case can be made either for or against global warming. However, 20 years may mean climate changes that cannot be undone. Government policy makers need to base their decisions on the most detailed information at hand, and they need to be able to make decisions now.

Digital measuring devices, satellites, and other methods of data-collection have eliminated much of the error and decreased the time taken to gather the required information, but these same technologies have been available for only the past 40 years. Most of our climate history has been recorded using older and less reliable methods. One new method of measuring ocean temperature is acoustic thermometry. This method uses

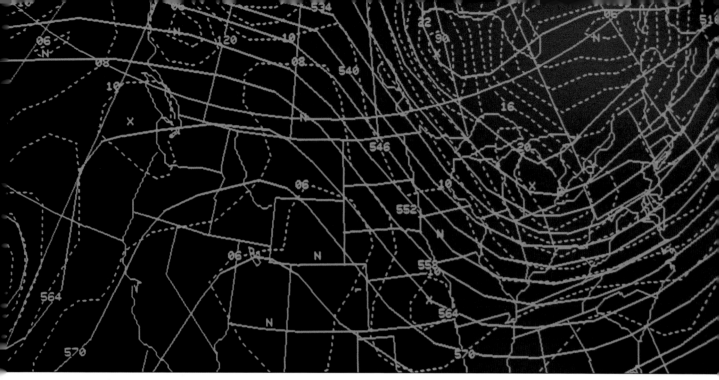

Weather maps are used to chart information gained from weather satellites. These maps help meteorologists predict weather patterns.

sound waves to measure the temperature of the ocean at various points and from a considerable distance. Perhaps new technologies and methods such as this one will increase scientists' ability to record and predict those factors that affect climate.

KEY CONCEPTS

Aerosols Aerosols are tiny particles in the air. Much of the aerosols in the air come from volcanoes that erupt and throw smoke and small particles into the air. Aerosols can also come from desert dust that is blown into the atmosphere by wind. These particles can have a big impact on short-term climate. In 1991, Mount Pinatubo in the Philippines erupted, shooting a great deal of aerosol into the atmosphere. As a result of this eruption, many countries recorded cool summers in 1992 and 1993.

Intergovernmental Panel on Climate Change (IPCC) In 1988, the United Nations Environment Programme (UNEP) and the World Meteorological Organization established IPCC to study the issue of climate change, study the possible effects of climate change on society and the environment, and create strategies to deal with the problem. IPCC's first report in 1990 served as a basis for the UN Framework Convention on Climate Change, presented in Rio de Janeiro, Brazil, in 1992. IPCC involves hundreds of scientists from all over the world.

Satellite weather determination Orbiting satellites above Earth can scan and gather massive amounts of information about weather. This information can be used to make much more accurate predictions about the daily or weekly weather for a region. In some cases, cloud formations can be analyzed to determine wind speeds. These satellites are especially useful for finding a tropical storm or other violent weather pattern before it can put people in danger. Others use these satellites to predict rainfall, cloud patterns and cycles, and any number of other weather phenomena.

From Ice Age to Hothouse

Most of the scientific community has accepted global warming as fact. IPCC is still convinced that Earth's temperature will continue to rise in the future.

One of the dangers of a warming trend is that ice above sea level will melt. If the planet warmed up enough to melt the polar ice caps over Greenland and Antarctica, sea levels would rise. This, in turn, would cause flooding in many coastal areas. Islands would be either partially or completely covered.

The United Nations Environment Programme (UNEP) reports that the sea level has risen between 4 and 10 inches (10–25 cm) in the past 100 years. Arctic pack ice and sea ice have been shrinking in this century, and the extent of the Arctic sea ice has declined by almost one-third in the past 130 years. Greenland has experienced melting. In

---

Carbon dioxide emissions may be causing global warming.

---

a positive feedback loop, the carbon dioxide and methane gases locked in the ice in these regions would be released into the atmosphere, causing even more global warming.

Antarctic ice, however, seems to be experiencing only small changes. There is warming in the Antarctic Peninsula and a reduction in ice shelf size in the northwest, but there is no other change in temperature or ice cover over the rest of Antarctica. A rise in global temperatures would most likely cause more snow to fall in the Antarctic. This would result in a buildup of aboveground water.

Since the 1992 Rio Earth Summit, the public has become more aware of the issue of global warming through classroom teaching, news reports, and research. This has led to greater pressure on politicians to control greenhouse gases.

Carbon dioxide is still seen as the major culprit in the warming trend. Since the Industrial Revolution, carbon dioxide concentrations in the atmosphere have increased from 280 **parts per million** (ppm) to about 370 ppm. A further rise to between 490 and 1,260 ppm is predicted. This is what IPCC points to as the cause of the predicted 2.5 to 10.4 °F (1.4–5.8 °C) increase in global temperatures. Yet, since 1992, the reliance on fossil fuels for energy has not decreased. Although a recent worldwide move to natural gas has reduced total emissions and improved power station efficiency, the United States, which produces 25 percent of the world's carbon dioxide, is still getting more than half its electricity from coal-fired power stations.

Using nuclear power stations instead of coal-fired power stations to generate

TUVALU

As Earth heats up, not only does ice melt and run into the sea (raising sea level), but water expands as a result of heating. This is a major problem for people living on low-lying land. The country of Tuvalu is a small island in the South Pacific and one of the lowest-lying countries in the world. In fact, no point on Tuvalu is higher than 15 feet (4.5 m) above sea level. The spring tides of 2000 reached 10.5 feet (3.2 m) higher than normal sea level, which threatened to put much of Tuvalu under water. If global warming continues and sea levels continue to rise, this tiny island may be swallowed by the ocean.

electricity could help the United States cut its carbon dioxide emissions, but the nuclear energy industry has been hurt by disasters such as Three Mile Island (1979) and Chernobyl (1986). However, with people calling for cleaner energy, the supporters of nuclear power have had their hopes raised. Spokespersons argue that nuclear power alone could provide the clean energy that is necessary for the future. Public opinion of nuclear energy is still low in most developed countries, but many developing countries see advantages in using nuclear energy. China, for example, has seven nuclear power stations in operation and four nuclear reactors under construction. Greenpeace warns that nuclear energy will involve a large investment for these countries, which will increase their debt. While nuclear power stations do not generate carbon dioxide, they do generate radioactive waste and weapons-usable materials, both of which are dangerous to the environment and expensive to eliminate.

Tuvalu, a small island nation in the South Pacific, has an average temperature of 86 °F (30 °C) and an annual rainfall of 139 inches (353 cm). Tuvalu is one of the lowest-lying countries in the world, and it is threatened by rising sea levels.

There has been talk about switching from nonrenewable energy sources to **renewable energy** sources such as wind, solar, and **hydroelectric energy**. Yet the technology needed to make these renewable forms of energy useful and productive is developing slowly. Renewable energy sources account for less than eight percent of total energy production in the United States and less than 10 percent globally. The International Energy Agency (IEA) states that by 2050 renewable energy sources could and should account for half of the energy used by humans. It has even entered into joint projects with the European Union (EU) to develop these technologies, but most energy research is still focused on traditional power generation. In 2002, for example, the U.S. government's budget for research and development of coal was $319 million, while only $248 million was being spent to research and develop renewable energy sources.

Political decisions are largely based on public pressure. The fossil fuel industry needs people to continue using its products in order to remain in operation. If there was a worldwide switch to renewable energy sources, all the jobs connected to the fossil fuel industry would be in jeopardy, and the machines and technologies would become obsolete. This is why the fossil fuel industry spends a great deal of money **lobbying** the government to continue its support of fossil fuels.

In the 1990s, lobby groups such as the Global Climate Coalition were set up by industry, especially the oil industry, to cast doubt on the idea that human activity was causing global warming. These groups warned the public about the economic damage that would be caused by the Rio and Kyoto agreements, both of which call for a reduction in greenhouse gas emissions. When evidence of global warming increased, and the public began thinking greener, many of the major oil companies, including Shell Oil

LARSEN B ICE SHELF

Antarctica is where 90 percent of Earth's ice lies. The Antarctic Peninsula has seen a 4.5 °F (2.5 °C) increase in temperature over the last 50 years, and the northwest ice shelves have been shrinking. For the most part, however, there has been no change in temperature or ice cover in Antarctica. One event that disturbed scientists was the rapid collapse of the Larsen B ice shelf in 2002. This ice shelf, which covered 1,255 square miles (3,250 sq km) and was 656 feet (200 m) deep, collapsed within 35 days and fragmented into small icebergs. Scientists had predicted its collapse as a result of local warming in the area, but no one expected the collapse to happen so quickly.

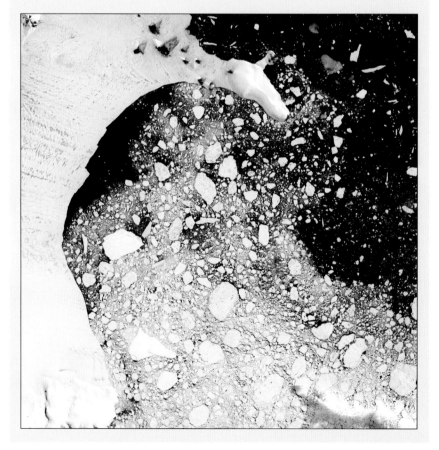

and British Petroleum, left the coalition. These companies began researching and investing heavily in renewable energy sources. One of the world's largest and richest oil companies stuck firmly to oil, coal, and natural gas. Although a target for protesters, they, and associated industries in general, maintain that the benefits of fossil fuels outweigh the costs. While many people are excited about renewable energy sources, those in nonrenewable industries see no reason to abandon worldwide economic growth because of a warming trend that may or may not be caused by human activities.

Some countries have been more active than others in making policies to cut GHGs.

The EU has long been dependent on fossil fuels, yet it appears to be the staunchest supporter of the Kyoto Protocol, which calls for a five percent reduction in carbon dioxide emissions worldwide by 2010.

Although many developing countries see the advantages of nuclear power, they are also in favor of greener methods of energy production. Many have no existing power plants for energy production, and there are good reasons why they should start with renewable energy sources. For example, at their present rate of use, oil and gas reserves may last between 40 and 70 years. Coal might have 200 years left. As developing countries become developed countries, their use

of fossil fuels will increase, and reserves will decrease faster. Renewable energy sources would meet the energy needs of these developing countries well into the future, while factories, mines, refineries, and plants that rely on fossil fuels may be useless in 100 years.

The U.S., on the other hand, has a strong investment in fossil fuels, and it would require more incentive for them to abandon technology and practices that already produce a great deal of energy at a reasonable price. Unfortunately, the absence of the United States from the energy revolution has had a negative effect on the international effort to counteract global warming.

Many countries have adopted air quality standards to protect the public from damaging pollutants, including sulfur dioxide and carbon monoxide.

Wind farms are clean sources of renewable energy. Prime sites for wind farms have average wind speeds greater than 17 miles (27 km) per hour. The most common wind turbines in commercial operation produce 600 kilowatts of power.

KEY CONCEPTS

Industrial Revolution The Industrial Revolution refers to a change in the way a country produces its goods and the way people live. Both in Europe and the U.S., the Industrial Revolution was made possible by many incredible inventions of the 1700s and 1800s. The electric battery, steam engine, coke smelter, cotton gin, and spinning jenny all made it possible to produce more products cheaper and faster. Hand tools and hand-made products were replaced by mass-produced items that more people could afford. It also involved a movement from country life to city life as people moved in search of factory jobs.

Nuclear disasters In April 1986, during a test at the Chernobyl nuclear reactor near Kiev in the former Soviet Union (now the Ukraine), uncontrolled nuclear reactions caused a series of explosions. As a result, the containment structure was ruptured, sending massive amounts of radiation into the atmosphere. This was the worst nuclear accident in history, and as of 2002, it had caused the deaths of more than 25,000 people who were exposed to high levels of radiation. Medical experts predict an increase of nearly 50,000 cancer-related deaths over a 50-year period as a result of fallout from the accident at Chernobyl. A similar but less severe accident occurred in 1979 at Three Mile Island in the United States. Although there was a radiation leak due to human error and equipment failure, no one was killed.

Born: March 21, 1768, in Auxerre, Bourgogne, France
Died: May 16, 1830, in Paris, France
Legacy: First to use the analogy of a greenhouse to explain why Earth is warmer than Mars

Navigate to the Web site **http://gaw. kishou.go.jp/wdcgg.html** for more information about greenhouse gases. Also click on **www-gap.dcs.st-and. ac.uk/~history/Mathematicians/ Fourier.html** to learn more about Jean-Baptiste Fourier.

People in Focus

Jean-Baptiste Fourier showed talent for literature in school, but by the age of 13, mathematics became his real interest. When he was 19, he decided to train for the priesthood and entered a Benedictine abbey. He did not take his vows and become a priest. Instead, he became a teacher at the Benedictine college where he had studied.

In 1793, Fourier became involved in politics and joined the local Revolutionary Committee. As a result of his involvement in politics, he was arrested, imprisoned, and released in 1794.

In 1798, Fourier joined Napoleon's army as a scientific adviser during its invasion of Egypt. During this time, he helped establish educational facilities in Egypt and carried out archaeological explorations.

Fourier's work on the mathematics of heat began around 1804, and by 1807, he had completed his important book *On the Propagation of Heat in Solid Bodies.*

In 1827, Fourier first used the analogy of a greenhouse trapping the sun's warmth to explain why Earth was not as cold as Mars. He suggested that heat from oceans and land masses reflected back into space was trapped by gases in the atmosphere. This creates a warm blanket around the planet that makes Earth fit for life.

Mapping Climate Change

Figure 1: Per capita emissions of carbon dioxide

- ■ High Emitters
- ■ Medium Emitters
- ■ Low Emitters
- ■ No Data

Scale 1:79,545,000

Charting Climate Change

Figure 2: Solar cycle and average temperature in the Northern Hemisphere

There is some evidence that solar activity affects global temperatures.

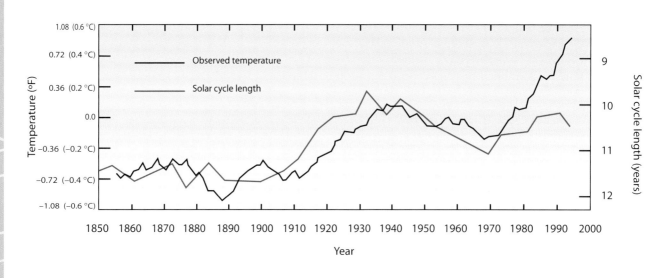

Figure 3: The greenhouse effect

Certain gases in the atmosphere trap heat and reflect it back to Earth.

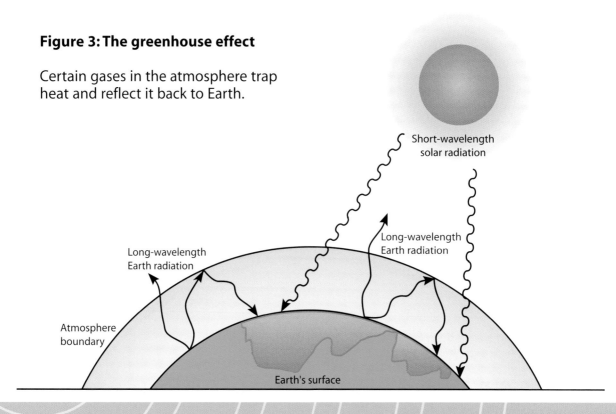

Figure 4: Global temperatures from 1850 to 1999 and projected estimates to 2100

Global temperatures are projected to rise over the next 100 years.

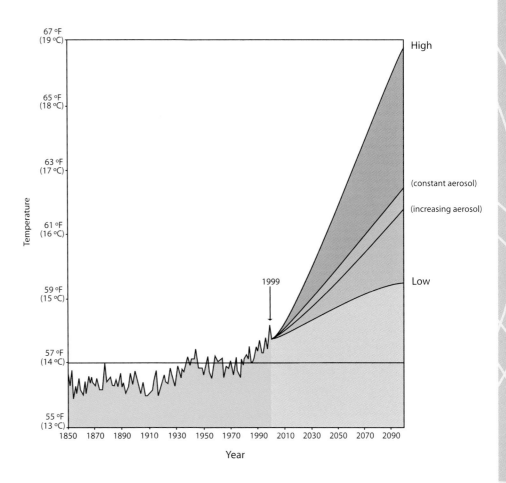

Figure 5: Global production of chlorofluorocarbons (CFCs)

The production of CFCs has dropped dramatically since the signing of the Montreal Protocol in 1987.

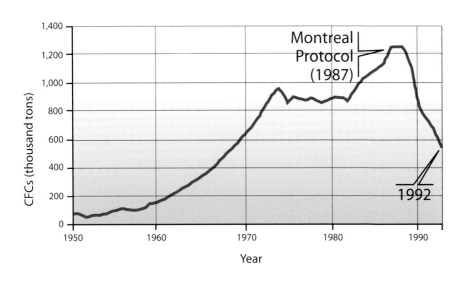

Climate and Civilization

Climate has always had a large impact on human history. During the ice ages, ice built up dramatically. This led to a drop in sea level and the formation of land bridges. These land bridges allowed humans to migrate to areas they would not have been able to reach previously without a ship. Global warming 7,000 years ago raised the sea to a level where the Black Sea and the Mediterranean Sea were joined. A rise in temperature around A.D. 1000 allowed Vikings to colonize northern lands that had been previously cut off by ice.

These are all examples of fluctuations in what has been a fairly stable 9,000 years. The Holocene Period has given humans time to gradually develop agricultural systems.

Agriculture supports a much greater number of people than hunting and gathering, so the human population has grown quickly. But agriculture depends on specific local conditions. If farmers in one area are growing wheat, they need at least 12 inches (30 cm) of precipitation per year to sustain their crop. If climate change brings much more annual rain to that area, it will no longer be suitable for the wheat crop. Obviously, crops that grow in hot and dry areas are unsuitable for cool, wet conditions, and vice versa. Because weather is so difficult to forecast, growers need to be able to rely on a stable climate to dictate what and when to plant. Agriculture is vulnerable to even small changes. A few weeks without rain can mean a lost crop. That means no food or income for farmers. In many ways, farming is much like gambling—the stakes are high, and much depends on the "luck of the draw." While other human activities can be adapted to suit weather conditions, agriculture remains sensitive to climate change.

Food production and other market-related elements of life on Earth should be able to adapt to global warming. Better ways to detect and predict weather could allow farmers to adapt by using new crops that can survive in a new climate. Industry and commerce should be able to continue without much difficulty. What about non-market elements such as natural ecosystems? How will

> **Earth's stable climate over the last 9,000 years has allowed humans time to develop agriculture systems.**

grasslands and ancient forests be able to adapt to rapid changes in temperature and rainfall? What will happen to animals that are suited for conditions that disappear? The arctic hare could not exist in a desert, nor could great reptiles survive a snowy winter.

Some areas would prosper from a change in climate, while others would suffer. Imagine a vast desert being converted to a productive, crop-bearing landscape in the course of several years or a grassland paradise being slowly converted to desert. How would farmers, deprived of useful land, be supported? Would water be shared or sold to those who had none? Human social structures and practices that have evolved in relation to weather patterns would also have to be considered.

IPCC has examined many of the possible impacts of global warming in order to determine which areas would be able to adapt easier than others. It concludes that the higher temperatures and rainfall that are expected in coming decades could have some unexpected advantages. Deaths from extreme cold would decrease, and fewer crops would be lost to early frost.

In spite of the handful of advantages, global warming would cause many serious problems. Older people, the urban poor, and livestock would suffer from more heat-related deaths and illnesses. There would be more damage from floods and landslides, including soil erosion. Diseases such as malaria would spread. Countries such as Bangladesh, Vietnam, and Egypt would be seriously affected by any rise in sea level.

Droughts occur when surface and subsurface water supplies are below normal. Droughts can have serious social, economic, and environmental impacts.

CLIMATE CHANGE AND THE YEARLY MONSOON

Much of Asia relies on yearly monsoons, or great rains, for the health of crops. If this seasonal rainfall is changed in location or amount, millions of people will be affected. Many will be thrown into poverty and starvation. Although global warming may mean more average global precipitation, no one knows where it will fall.

A sea level rise of three feet (1 m) would reduce Bangladesh's land area by one-fifth and displace millions of people.

In November 2001, *The Ecologist* published a special issue on climate change. Its warning about the effects of global warming is more severe than the warning issued by IPCC. It points out that 10 trillion tons (9 trillion t) of methane are locked in crystals in the permafrost and on the ocean floor, and this methane could be released by global warming. Such an event would be a severe example of positive feedback and further increase global warming. The magazine concludes that the most important issue is survival, not development. The strategies for survival are different from the strategies for growth, and many reports, including the *The Ecologist*, suggest that in order to survive global warming people need to make drastic changes.

Instead of debating whether global warming is occurring, people may do well to assume that global warming is a fact and that the survival of the human race is more important than economic growth.

According to the United Nations Environment Programme (UNEP), a vegetarian meal with locally produced ingredients generates 0.44 pounds (0.2 kg) of carbon dioxide. Compare this with a meal of imported pork and vegetables, which generates four pounds (1.8 kg) of carbon dioxide. According to these figures, the risk of global warming could be reduced if everyone became vegetarian and ate local produce. However, this would hurt those who are involved in the meat or food distribution industry.

It is known that burning fossil fuels puts more GHGs into the atmosphere. It would help if people did not drive, use air-conditioning, heat swimming pools, or mow their lawns. In fact, cutting down on the use of electricity would help, too. A summary of this argument might be simply this: The way to prevent global catastrophe is to use renewable energy, reduce wasteful consumerism, and oppose big business.

Other people do not feel that global warming is the most important problem facing the world. In the 2001 book, *The Skeptical Environmentalist*, Bjørn Lomborg claims that the world is spending too much time and money on a global warming that may be smaller than expected. He argues that "what matters is making developed countries rich and giving citizens of developed countries greater opportunities." For people such as Lomborg, economic growth should not be sacrificed for an event that may or may not happen.

FLOODING IN BANGLADESH

Flooding is common in Bangladesh, and a major rise in sea level would put much of this country, and others, under water indefinitely. In July 2002, flooding in South Asia killed more than 500 people in Bangladesh, Nepal, and India and left millions homeless or stranded.

Flooding has become a major problem for low-lying countries such as Bangladesh. Some experts believe global warming is the cause of this flooding.

In order to enforce pollution standards, pollution control authorities measure pollutants in the atmosphere and at specific sources, such as industrial smokestacks.

KEY CONCEPTS

Drought Drought is a period of dryness that can cause extensive damage to plants. In 2002, drought affected most of the western U.S., 90 percent of New South Wales in Australia, the horn of Africa, Afghanistan, and parts of South America. Droughts can cause serious problems. For example, the drought in Australia may cause an economic recession, and as a result of the drought in Africa, six million people are in need of food aid. Some of the droughts of 2002 were the worst in 100 years.

Monsoon The term "monsoon" is used to describe seasonal winds or the rain brought by these winds. In South Asia, monsoon winds bring heavy rain during the summer and autumn. Much of the agriculture in India and Southeast Asia depends on the monsoon rains. Global warming is expected to increase the amount of rainfall worldwide, but there are already concerns because the yearly monsoon rains have not been as heavy in South Asia as in years past.

The Kyoto Process

In 1992, most of the countries of the world came together in Rio de Janeiro to show that global concerns were as important as local concerns. The meeting was a historic one because it showed true global cooperation.

The same type of global cooperation resulted in the 1987 Montreal Protocol. Evidence had shown that chlorofluorocarbons (CFCs) were destroying Earth's ozone layer, which protects the planet from UV radiation. Seeing the risk of worldwide UV increases, the world community gathered to reduce CFCs. In fact,

they agreed to systematically ban and find substitutes for chemicals that depleted the ozone layer. There were even procedures to help developing countries achieve these goals without slowing their growth and development. This was a truly cooperative process that has worked. Spray cans that used to contain CFCs have "no CFCs" labels on them now. New chemicals have been discovered that can be used as coolants and solvents. Whether or not the

The Kyoto Protocol calls for carbon dioxide emissions to be cut by five percent.

ozone hole continues to cause problems in the future, the world community did take swift action to deal with this known threat to the ozone layer.

The 1992 Rio Earth Summit dealt with issues that were larger than those covered by the Montreal Protocol—worldwide energy production and severe climate change. The 1992 Climate Convention, which has been ratified by more than 170 countries, urged participating countries to reduce carbon dioxide emissions to 1990 levels by 2000. Developing countries were allowed to opt out of this commitment. They were asked, however, to keep track of their carbon dioxide emissions because GHG emissions are expected to grow as these countries develop. China's carbon dioxide emissions could double soon. On the other hand, Germany, a

▨ **Because of environmental issues, many countries require mining companies to secure government permits before they mine for coal.**

member of the European Union (EU), set itself a goal of 30 percent reduction by 2005. This first step in the fight against global warming can be seen in many ways—a very small band-aid solution for a huge problem; an overreaction to a problem that may not exist or may not be fixable; or an opportunity to develop new technologies that do not emit GHGs.

The Kyoto Protocol process began in Kyoto, Japan, in late 1997 and went further than the 1992 Climate Convention. The developed countries that signed the agreement decided to work together to reduce overall emissions by five percent below 1990 levels by 2008 to 2012. This would be done by assigning specific emission targets to countries. EU countries would cut carbon dioxide emissions by eight percent; the U.S. would cut emissions by seven percent; and Canada would cut emissions by six percent. Some countries, such as New Zealand, Russia, and the Ukraine, would not be required to reduce carbon dioxide emissions. Australia and Iceland were even allowed to increase carbon dioxide emissions. A global reduction of five percent would not be enough to stop a global warming process (that would require at least a 60 percent reduction), but it would be a start. It would also encourage countries to begin research and development of technology that allowed further cuts to greenhouse gas emissions.

The participation of the United States in Kyoto was considered critical because the U.S. emits more GHGs than any other country. However, as the signing date approached in 2001, the U.S., under President George W. Bush, pulled out of the Protocol. The other countries, led by the European Union, decided to sign the Protocol. To achieve legal force, the Protocol had to be signed by 55 countries. Also, 55 percent of all emissions had to come from signing developed countries. As of July 2002, the Protocol had been ratified by 77 countries, but the U.S., which accounts for 36 percent of the carbon dioxide emissions produced by developed countries, had not signed the Protocol. Without U.S. ratification, almost every other developed country will have to accept the accord to put the Kyoto Protocol into action.

Throughout the process of negotiation, the U.S. government was cautious about agreeing to a

▨▨▨▨ **Most air pollution is caused by the burning of fossil fuels to power industry and motor vehicles.**

process that might hurt its economy. Former Vice President Al Gore said that the U.S. was "perfectly prepared to walk away from an agreement that we don't think will work." Cutting down on fossil fuel use is seen by many decision makers in the U.S. as a near impossibility because of how much the country relies on it already. They say the cost of changing over to alternatives would be high and hurt the gross domestic product (GDP). The EU pushed for Kyoto to require a 15 percent reduction for developed countries. It still hopes to meet this mark in its part of the world and has been using strategies such as carbon taxes to reduce emissions. The EU can reach its Kyoto targets with only a 0.06 percent reduction in GDP.

HOLE IN THE OZONE

The ozone layer in the atmosphere stops most solar ultraviolet radiation from reaching Earth. A hole in the ozone over Antarctica was first noticed in the 1970s by a research group from the British Antarctic Survey. The loss of ozone over the Antarctic was first measured in the 1980s. When the first measurements were taken, the drop in ozone levels in the stratosphere was so dramatic that the scientists thought their instruments were wrong.

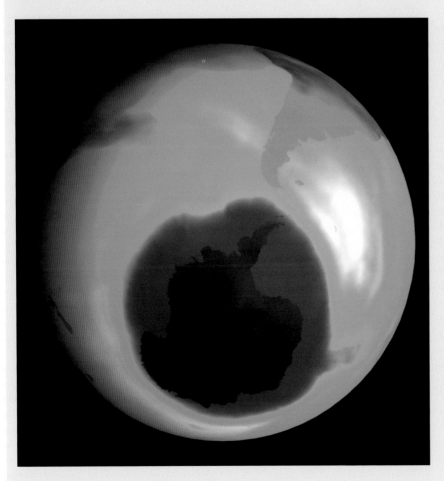

Over the years, the hole in the ozone has increased rapidly, and it is currently as large as the Antarctic continent (1.5 times larger than the United States). The hole lasts for only two months, but its timing could not be worse. Just as sunlight reaches dormant plants and animals begin to return and bear young, Antarctica is hit with a harmful dose of ultraviolet radiation that can cause skin cancer, injure eyes, harm the immune system, and upset the fragile balance of an entire ecosystem. After eight weeks, the hole leaves Antarctica and passes over more populated areas, including the Falkland Islands, South Georgia, and the tip of South America.

It was a major shock to the world community to hear that chlorofluorocarbons (CFCs) were destroying the ozone. People became informed about aerosol sprays, chlorofluorocarbons, and skin cancer. In 1987, the Montreal Protocol helped to ban the use of CFCs. With that decision, people may have averted a disaster they would have brought on themselves.

Japan would actually see a 0.9 percent boost in its GDP.

The U.S. has even refused to abide by the commitments it made in the early 1990s. It has not brought emissions to 1990 levels, and by 1996, its emissions had increased nine percent over 1990 levels. If the trend continues, the U.S. will have increased emissions by 54 percent over 1990 levels by 2020.

When George W. Bush came to power, U.S. national policy was reviewed. The U.S. National Academy of Sciences was asked to research the science behind Kyoto. The Academy largely agreed with IPCC's assessment of the danger from global warming, but it added that it could not rule out that a significant part of global warming is due to natural causes.

Although computer models are improving, the global climate system is too complex to be simulated, and predictions can only be as good as the data, which is often inaccurate. Still, the majority of the world's scientific community largely supports the Kyoto Protocol and the need to cut down on the amount of greenhouse gases being released into the atmosphere.

Raw sewage, industrial wastes, and agricultural chemicals are the main causes of water pollution.

KEY CONCEPTS

Chlorofluorocarbons (CFCs)
CFCs were seen as a group of miracle chemicals when they were first produced. They could be used to cool, to clean, and to propel aerosols from cans safely. They were stable chemicals, non-toxic to humans, and supposedly environmentally neutral. However, researchers discovered that as the CFC molecule broke down in the atmosphere parts of it affected the ozone and created a hole that let in dangerous ultraviolet radiation. By the time of the discovery, CFCs were in such wide use in cars, refrigerators, and household products that it took a worldwide effort and great expense to halt their production.

Gross domestic product (GDP)
The gross domestic product is the total value of goods produced and services provided in a country in one year. GDP is used to measure the wealth of a country. Some economists argue that GDP does not tell the whole story. For example, GDP measures the amount of lumber sold, but it does not account for the environmental damage caused by logging.

Born: February 19, 1859, in Wijk, Sweden

Died: October 2, 1927, in Stockholm, Sweden

Legacy: Made important calculations describing the effect of atmospheric carbon dioxide on global temperature

Navigate to the Web site **www.nobel.se/chemistry/laureates/1903** for more information about Svante Arrhenius. Also click on **www.tamug.edu/labb/ Global_Warming_Info.htm** to learn more about the effect of carbon dioxide emissions on global temperatures.

People in Focus

More than 100 years ago, Swedish scientist Svante Arrhenius wondered whether the temperature of the ground was in any way influenced by the presence of the heat-absorbing gases in the atmosphere. Drawn to the topic by Jean-Baptiste Fourier's work on heat in solid bodies, he began to ponder the potential effects of carbon dioxide and other gases on Earth's atmosphere. The increase in fossil fuel consumption, he believed, might well contribute to the natural global warming caused by the sun.

On December 24, 1894, he rolled up his sleeves and began what he later described as the most tedious calculations of his life. He was the first person to investigate the effect that doubling atmospheric carbon dioxide would have on global climate. Arrhenius determined that a doubling of carbon dioxide in the atmosphere would lead to an average temperature increase of 9 to 11 °F (5–6 °C).

In the paper Arrhenius presented to the Swedish Academy, he was not alarmed at the future climate change. He saw the warming of the cold North he lived in as a positive change. Besides, Arrhenius figured it would take 3,000 years to double atmospheric carbon dioxide and that fossil fuels would be used up by then. The global warming question has been debated ever since and is still a main concern of scientists today.

In 1903, Arrhenius was awarded the Nobel Prize for Chemistry.

Climate and Political Will

While global warming is largely accepted as fact, even by most of the oil industry, not everyone is in favor of the Kyoto Protocol solution. Some say the carbon dioxide reduction targets in the Kyoto Protocol will cost too many jobs and too much money. Others say the reduction targets are unfairly distributed. Even though the U.S. administration under former president Bill Clinton promised a seven percent reduction in carbon dioxide emissions, the Bush administration has stated that a seven percent reduction would cripple economic growth in the United States. Governments of other rich countries were also worried that major cuts in GHGs would mean lost jobs. The Bush administration argued that because developing countries were emitting more and more carbon dioxide they should be given reduction targets.

Carbon dioxide emissions are rising very rapidly in developing countries. As a country becomes more developed, more electricity, cars, and buses are used. These countries argue that while the goals of Kyoto are desirable, emission cuts would hinder their growth. The Protocol gave a grace period to these developing countries because they had produced very little of the GHGs before 1990. Also, the global community does not want to slow down the development of these countries.

The Kyoto Protocol was designed with special rules and regulations to deal with arguments such as those previously mentioned. First, countries would be able to use their emission permits to trade with other countries. If a country cuts emissions more than it is required, it is given credits that can be sold to countries that cannot meet their target. This would mean a great deal of money for countries such as Russia and the Ukraine, which have reduced emissions greatly since 1990, mainly as a result of the breakup of the Soviet Union. That money could be used to rebuild these countries. On the other hand, a country that is very prosperous, such as the U.S., would have to buy credits from other countries in order to continue its rapid growth. By 2010, the emissions trade market could range anywhere from $10 billion to $3 trillion. Credits would also be awarded to countries that achieved their emissions cuts.

Hybrid engines use gasoline and electricity to power motor vehicles, which improves fuel efficiency and reduces emissions.

CARBON DIOXIDE EMISSIONS

The figures for carbon dioxide emissions from fossil fuels show how little was achieved in reducing the buildup of greenhouse gases during the 1990s. Instead of stabilizing emissions at 1990 levels, as called for by the 1992 Climate Change Convention, the U.S. increased its emissions by 16 percent. The EU held carbon dioxide emissions steady, while China was able to hold its increase of emissions to 21 percent, despite tripling its economic output since 1990. Eastern Europe's post-Soviet economic collapse is the main reason for its reduced emissions. Emissions from the developing world are also on the rise.

CARBON DIOXIDE EMISSIONS (MILLION TONS OF CARBON EQUIVALENT)

YEAR	USA	WESTERN EUROPE	EASTERN EUROPE	CHINA	JAPAN	REST OF WORLD	WORLD TOTAL
1991	1,480 MtC (1,343 t)	1,100 MtC (1003 t)	1,312 MtC (1,190 t)	712 MtC (646 t)	309 MtC (280 t)	1565 MtC (1420 t)	6,484 MtC (5,882 t)
1992	1,505 MtC (1,366 t)	1,064 MtC (965 t)	1,239 MtC (1,124 t)	736 MtC (668 t)	315 MtC (286 t)	1621 MtC (1471 t)	6,481 MtC (5,880 t)
1993	1,535 MtC (1,393 t)	1,056 MtC (958 t)	1,140 MtC (1,034 t)	785 MtC (712 t)	312 MtC (283 t)	1718 MtC (1559 t)	6,547 MtC (5,939 t)
1994	1,562 MtC (1,417 t)	1,050 MtC (953 t)	1,007 MtC (914 t)	847 MtC (768 t)	330 MtC (299 t)	1776 MtC (1611 t)	6,572 MtC (5,962 t)
1995	1,576 MtC (1,430 t)	1,073 MtC (973 t)	967 MtC (877 t)	869 MtC (788 t)	328 MtC (298 t)	1883 MtC (1708 t)	6,695 MtC (6,074 t)
1996	1,632 MtC (1,481 t)	1,102 MtC (1,000 t)	949 MtC (861 t)	885 MtC (803 t)	340 MtC (308 t)	1940 MtC (1760 t)	6,849 MtC (6,213 t)
1997	1,656 MtC (1,502 t)	1,109 MtC (1,006 t)	888 MtC (806 t)	908 MtC (824 t)	341 MtC (309 t)	2015 MtC (1828 t)	6,917 MtC (6,275 t)
1998	1,658 MtC (1,504 t)	1,109 MtC (1,006 t)	871 MtC (790 t)	887 MtC (805 t)	331 MtC (300 t)	2038 MtC (1849 t)	6,894 MtC (6,254 t)
1999	1,682 MtC (1,526 t)	1,085 MtC (984 t)	908 MtC (824 t)	873 MtC (792 t)	338 MtC (307 t)	2083 MtC (1890 t)	6,970 MtC (6,323 t)
2000	1,732 MtC (1,571 t)	1,102 MtC (1,000 t)	930 MtC (844 t)	854 MtC (775 t)	346 MtC (314 t)	2137 MtC (1939 t)	7,102 MtC (6,443 t)

MtC = Million tons of carbon equivalent   t = metric tonnes

Finally, developed countries could win credits by helping developing countries cut their carbon dioxide emissions.

Clearly there is a lot of money at stake, and money leads to further disagreement.

What if a country not only cuts its carbon dioxide output, but actually removes carbon dioxide from the atmosphere? Russia and Canada have two of the largest forests in the world. A forest is a **carbon sink**, a place where carbon dioxide is literally sucked out of the air. These two countries could cut down the forests for profit and economic growth. Should they receive credits for not cutting down their forests? Russia thinks so.

Canada would like credits for supplying the U.S. with clean energy (derived from hydroelectric generating plants), and it has threatened to pull out of the Protocol if it does not receive those credits. Canada believes it is helping the U.S. reduce its need for more fossil fuels. Complex arguments such as these lead many to believe that Kyoto is doomed to fail.

Many of the actions required to reduce the emission of carbon dioxide are easy to carry out, even at a local level. Generating plants that run on fossil fuels may be around for a very long time, but the use of less energy and more energy-efficient appliances will reduce the amount of energy being produced. It can be as simple as turning off appliances when they are not in use. On a larger scale, companies can follow energy conservation strategies. There is approximately 68 billion square feet (21 billion sq m) of commercial space in the U.S., which annually consumes 5.8 quadrillion **BTUs** of energy at a cost of $70 billion. Commercial buildings account for about 15 percent of U.S. greenhouse gas emissions. The Energy Star Buildings Program, which was created by the Environmental Protection Agency (EPA), asks that building owners make upgrades to increase energy efficiency. In 2002, the top energy performing buildings in the U.S. used 40 percent less energy than the average building.

There are many ways to reduce greenhouse gas emissions.

Transportation is another area where emissions can easily be cut. Many people are beginning to take public transportation, and quite a few cities are increasing and improving their public transport services to accommodate this new trend. People who wish to have personal vehicles can now choose ones with hybrid gas/electric engines. These vehicles use much less fuel and help reduce GHG emissions.

Finally, industries can cut emissions by installing better heating and cooling systems and better insulation in factories and offices. Industries could also cut emissions by lowering production, but that kind of talk makes businesses nervous. Industry is based on production. However, there is only need for production if people continue to consume resources and products at a high rate. A reduction in personal consumption would cause industry rates to fall to a level that might achieve the targets of Kyoto.

World Energy Outlook 2001, a report from the International Energy Agency, points out that there is no shortage of global energy supplies. The International Energy Agency believes fossil fuels will continue to be the world's major energy source for the next two decades at least. It states that after 2020

KEY CONCEPTS

Environmental Protection Agency (EPA) The EPA is a branch of the U.S. government. It monitors environmental conditions in the U.S. and around the world. It produces educational programs for children and encourages the public to be more environmentally conscious and active. The EPA created the Energy Star Program, which rates the energy consumption and the amount of pollution of a variety of products. This helps the public make more environmentally friendly purchases. It also awards grants for projects that will lead to a healthier environment.

Hybrid engines These new types of engines are a combination of an electric battery and a standard fuel-burning engine. Some of the power comes from the electric engine. At other times, such as during acceleration, the conventional engine provides power. At the same time, the gas-burning engine is recharging the electric battery. This kind of system will allow some cars to achieve between 50 and 80 miles (80 and 129 km) per gallon. That kind of fuel efficiency would help reduce carbon dioxide emissions.

new technologies such as hydrogen-based fuel cells may provide cleaner energy. Whether or not humans are causing the present global warming, energy efficiency and clean air are worthy goals, and there are many ways to reduce greenhouse gas emissions with little or no cost to society.

ESTIMATES OF POTENTIAL GREENHOUSE GAS EMISSION REDUCTIONS IN 2020

SECTOR	EMISSIONS 1990 (MtC/YR)	POTENTIAL REDUCTIONS 2020 (MtC/YR)	NET DIRECT COSTS PER TON OF CARBON AVOIDED ($/tC)
Buildings (CO$_2$ only)	1,819 MtC/yr (1,650 t/yr)	1,102–1,213 MtC/yr (1,000–1,213 t/yr)	Most reductions available at negative direct costs
Transport (CO$_2$ only)	1,190 MtC/yr (1,080 t/yr)	331–772 MtC/yr (300–700 t/yr)	Probably less than $25/tC
Industry (CO$_2$ only)	2,535 MtC/yr (2,300 t/yr)	772–992 MtC/yr (700–900 t/yr)	More than half available at net negative direct costs
Agriculture (CO$_2$ only)	231 MtC/yr (210 t/yr)		Cost range $0–100/tC in most cases
Methane etc.	1,378–3,086 MtC/yr (1,250–3,086 t/yr)	386–827 MtC/yr (350–750 t/yr)	
Waste Methane only	265 MtC/yr (240 t/yr)	220 MtC/yr (200 t/yr)	About 75 percent of savings as methane recovery from landfill sites at net negative direct costs; 25 percent at $20/tC
Energy Supply/ Conversion	1,786 MtC/yr inc. above (1,620 t/yr inc. above)	386–772 MtC/yr (350–700 t/yr)	Many options available for less than $100/tC
Total	**7,418–9,126 MtC/yr (6,730–8,566 t/yr)**	**3,197–4,796 MtC/yr (2,900–4,463 t/yr)**	

Because of global economic development, IPCC estimates a range of emissions from 13,228 to 17,637 million tons (12,000–16,000 million t) of carbon equivalent (MtC) for 2020, almost double the 1990 figures. Even if reductions of 5,511 million tons (5,000 million t) of carbon equivalent could be achieved, greenhouse gas emissions in 2020 will still be above the 1990 level. About half of the potential for reducing emissions lies in hundreds of new technologies for energy efficiency in buildings, transport, and manufacturing industries (e.g., better insulation, more efficient heating and lighting, use of hybrid engines, fuel cells etc.). Note the range of uncertainty and the limited options for reducing greenhouse gas emissions in agriculture.

MtC = Million tons of carbon equivalent tC = ton of carbon equivalent t = metric tonnes of carbon equivalent

Concept Web

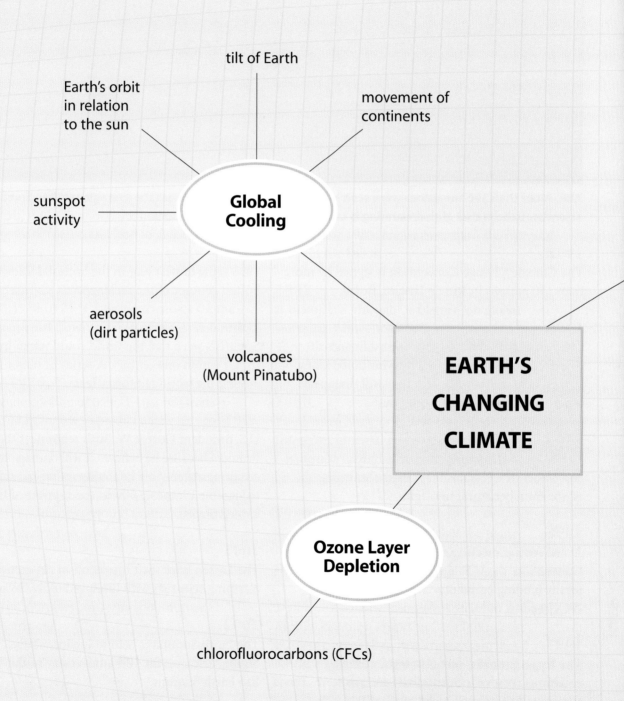

tilt of Earth

Earth's orbit
in relation
to the sun

movement of
continents

sunspot
activity

**Global
Cooling**

aerosols
(dirt particles)

volcanoes
(Mount Pinatubo)

**EARTH'S
CHANGING
CLIMATE**

**Ozone Layer
Depletion**

chlorofluorocarbons (CFCs)

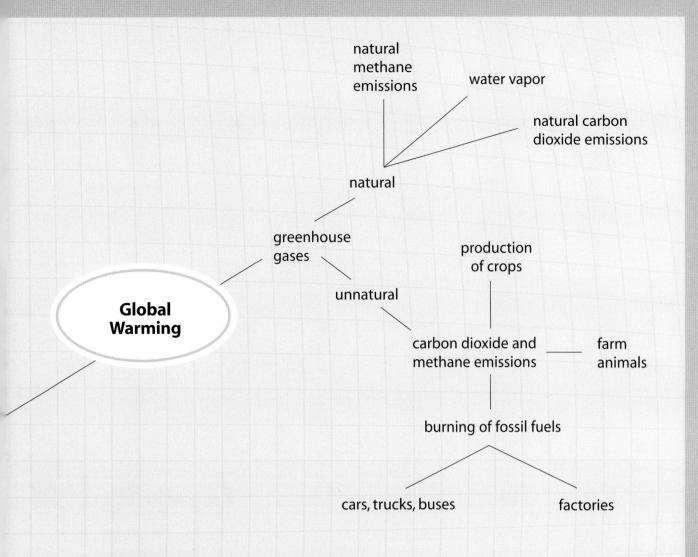

natural methane emissions

water vapor

natural carbon dioxide emissions

natural

greenhouse gases

Global Warming

production of crops

unnatural

carbon dioxide and methane emissions

farm animals

burning of fossil fuels

cars, trucks, buses

factories

MAKE YOUR OWN CONCEPT WEB

A concept web is a useful summary tool. It can also be used to plan your research or help you write an essay or report. To make your own concept map, follow the steps below:

- You will need a large piece of unlined paper and a pencil.
- First, read through your source material, such as *Earth's Changing Climate* in the Understanding Global Issues series.
- Write the main idea, or concept, in large letters in the center of the page.
- On a sheet of lined paper, jot down all words, phrases, or lists that you know are connected with the concept. Try to do this from memory.
- Look at your list. Can you group your words and phrases in certain topics or themes? Connect the different topics with lines to the center, or to other "branches."
- Critique your concept web. Ask questions about the material on your concept web: Does it all make sense? Are all the links shown? Could there be other ways of looking at it? Is anything missing?
- What more do you need to find out? Develop questions for those areas you are still unsure about or where information is missing. Use these questions as a basis for further research.

Quiz

Multiple Choice

1. Which of the following is NOT a greenhouse gas?
   a) water vapor
   b) nitrogen (N)
   c) carbon dioxide (CO_2)
   d) methane (CH_4)

2. The only fossil fuel below is:
   a) coal
   b) hydroelectric
   c) plutonium
   d) hydrogen fuel cells

3. Renewable energy sources include all of the following except:
   a) wind
   b) solar
   c) hydroelectric
   d) natural gas

4. In the last 100 years, Earth's temperature has risen about:
   a) 1 °F (0.6 °C)
   b) 2 °F (1.2 °C)
   c) 3 °F (1.8 °C)
   d) 5 °F (3.0 °C)

5. To stop the effects of global warming from carbon dioxide, humans would have to cut emissions by:
   a) 7 percent
   b) 5 percent
   c) 60 percent
   d) 25 percent

Where Did It Happen?

1. In 1987, world leaders gathered here to deal with the ozone problem.
2. In 2002, a huge ice shelf called Larsen B broke off here.
3. The first meeting to deal with global warming took place here in 1992.
4. This country produces 25 percent of Earth's carbon dioxide.
5. This developing country had seven nuclear power stations by 2002.

True or False

1. The greenhouse effect is a natural process.
2. Antarctica is losing ice rapidly throughout the entire continent.
3. Sea levels will rise if ice caps melt.
4. The U.S. has not signed the Kyoto Protocol.
5. In the last 9,000 years, the climate has changed dramatically.

Answers on page 53

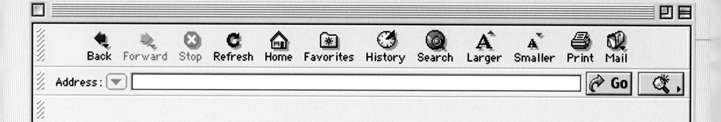

Internet Resources

The following Web sites provide more information about climate change:

THE UNITED STATES GOVERNMENT ENVIRONMENTAL PROTECTION AGENCY (EPA)
http://www.epa.gov/globalwarming
EPA's Global Warming Web site is provided as a public service. This Web site strives to present accurate information about the very broad issue of climate change and global warming in a way that is accessible and meaningful to all ages.

THE INTERGOVERNMENTAL PANEL ON CLIMATE CHANGE (IPCC)
http://www.ipcc.ch
IPCC assesses the scientific, technical, and socio-economic information that relates to human-induced climate change. IPCC's official Web site includes complete documents explaining all facets of global climate change.

THE PEW CENTER ON GLOBAL CLIMATE CHANGE
http://www.pewclimate.org
The Pew Center on Global Climate Change is a nonprofit, nonpartisan organization dedicated to providing credible information, straight answers, and innovative solutions to global climate change. This Web site contains the latest information about global climate change.

Some Web sites stay current longer than others. To find other Web sites that deal with climate change, enter terms such as "climate change," "global warming," or "carbon dioxide emissions" into a search engine.

Further Reading

Burroughs, William James. *Climate Change: A Multidisciplinary Approach.* New York: Cambridge University Press, 2001.

Drake, Frances. *Global Warming: The Science of Climate Change.* London: Arnold, 2000.

Leggett, Jeremy. *The Carbon War.* New York: Penguin, 1999.

Lomborg, Bjørn. *The Skeptical Environmentalist: Measuring the Real State of the World.* New York: Cambridge University Press, 2001.

Mendlesohn, Robert. *The Impact of Climate Change on the United States.* New York: Cambridge University Press, 1999.

Watson, Robert T., and the Core Writing Team, IPCC. *Climate Change 2001: Synthesis Report.* New York: Cambridge University Press, 2002.

Answers

Multiple Choice
   1. b)   2. a)   3. d)   4. a)   5. c)

Where Did It Happen?
   1. Montreal   2. Antarctica   3. Rio de Janeiro   4. United States   5. China

True or False
   1. T   2. F   3. T   4. T   5. F

Glossary

aerosols: dust particles in the atmosphere

atmosphere: the gases that surround Earth

biomes: large areas with similar vegetation

BTU (British thermal unit): the amount of heat needed to raise the temperature of one pound of water by 1.0 °F (0.6 °C)

carbon dioxide (CO_2): a colorless, odorless, incombustible gas that is formed during respiration, combustion, and organic decomposition and used in food refrigeration, carbonated beverages, and fire extinguishers

carbon sink: an area, such as a forest, where carbon dioxide is pulled out of the atmosphere

developed countries: countries in the industrialized world; highly developed economically and technologically

developing countries: countries that are undergoing the process of industrialization

emission: something that is put out

feedback: a response to a change

fluctuation: a change from normal conditions

fossil fuels: fuels such as coal, oil, and natural gas

global warming: an increase in the average temperature of the Earth's atmosphere, enough to cause climate change

greenhouse gases: atmospheric gases that can reflect heat back to Earth

Gulf Stream: a warm ocean current of the northern Atlantic Ocean off eastern North America

hydroelectric energy: electricity generated by turbines turned by running water

interglacials: warm periods between ice ages

lobbying: persuading someone that a particular thing should or should not happen

methane: an odorless, colorless, flammable gas that is the major component of natural gas

oceanographers: scientists who study the physical and biological aspects of the seas

parts per million: the number of one kind of molecule or particle in a random sample of one million molecules or particles

plankton: small or microscopic organisms that float or drift in great numbers in fresh or salt water and serve as food for fish and other large animals

renewable energy: energy obtained from sources that cannot be depleted

weather: the state of the atmosphere at a given time and place, accounting for variables such as temperature, moisture, wind velocity, and barometric pressure

Index

Index

Photo Credits

Cover: Gust front from a severe storm, Texas (**Corbis Corporation**); **Brian Cannon/Tuvalu Online**: page 23; **Corbis Corporation**: pages 2/3, 9, 10, 17, 21, 47; **Corel Corporation**: pages 6/7, 18; **Digital Vision Ltd.**: pages 1, 14, 25, 26, 34, 35, 40; **Mike Grandmaison**: page 38; **Warren Gretz/DOE/NREL**: page 42; **Ann Hawthorne/B&C Alexander**: page 11; **Landsat 7 Science Team & NASA GSFC**: page 24; **MaXx Images**: pages 36/37; **The Nobel Foundation**: page 41; **PhotoSpin Inc.**: page 15; **Science Photo Library**: page 27; **Greg Shirah, GSFC Scientific Visualization Studio**: page 39; **Tom Stack & Assoc.**: pages 12 (**Chip & Jill Isenhart**), 4 (**Mark Newman**); **UN/DPI**: page 32.